Food Chains

Theresa Greenaway

RAINTREE
STECK-VAUGHN
PUBLISHERS

A Harcourt Company

Austin   New York
www.steck-vaughn.com

CYCLES IN NATURE
Other titles in this series:
Plant Life

The Water Cycle

Cover: A grizzly bear eating a fish (*main image*); (*clockwise from top right*) a colorful variety of vegetables; a diagram of a food chain; spikes of ripening wheat; a cow grazing on grass.

Title page: A grizzly bear eating a fish

Contents page: A colorful variety of vegetables

Published by Raintree Steck-Vaughn Publishers,
an imprint of Steck-Vaughn Company

Library of Congress Cataloging-in-Publication Data
Greenaway, Theresa.
Food chains / Theresa Greenaway.
    p.   cm.—(Cycles in nature)
    Includes bibliographical references (p.).
    Summary: Explains how energy is transferred from plants to animals via the food chain.
    ISBN 0-7398-2730-8
    1. Food chains (Ecology)—Juvenile literature.
    [1. Food chains (Ecology). 2. Ecology.]
    I. Title.
    QH541.14.G72 2000
    577'.16—dc21   00-020920

Printed in Italy. Bound in the United States.
1 2 3 4 5 6 7 8 9 0 05 04 03 02 01

Picture acknowledgments
The publishers would like to thank the following for allowing their images to be used in this book: Bruce Coleman 18 (center)/M.P.L. Fogden, 23 (top)/Andy Purcell, 23 (bottom)/Kim Taylor, 24/Christer Fredriksson; Greg Evans 17; NHPA 16/Bryan & Cherry Alexander, 18 (bottom)/G.I. Bernard, 19/B. Jones & M. Shimlock, 27/Yves Lanceau; OSF *cover* (bottom left)/Niall Benvie, 12/Daniel J. Cox, 20/Mike Hill, 25/Mark Hamblin, 29/Bob Gibbons; Papilio 7 (top); RSPCA *cover* (top left)/Angela Hampton, 26/E.A. Jones; Still Pictures 4/Jean F. Stoick; Stock Market 5/Daniel Aubry; Tony Stone Images *cover* (main image) & *title page*/James Balog, 8/Nicholas Parfitt, 9/Daniel J. Cox, 11/Mark Petersen, 21/Kathy Bushue, 28/Darryl Tucker; Wayland Picture Library *cover* (top right), *contents page* & 7 (bottom). Artwork on pages 6, 10, 13, and 14–15 is by Peter Bull. Artwork on page 22 is from Wayland Picture Library. Border and activity box artwork is by Jan Sterling.

Contents

Eating to Live

Every living thing is eaten by something else. Dead or alive, plants and animals are all food for something. Even wood, bones, and fallen autumn leaves are eaten. The different kinds of living organisms in the wild are all dependent on one another. A hare may nibble on some leaves. Later, a lynx may catch and eat the hare. The process by which leaves are eaten by a hare, which is then eaten by a lynx, is an example of a food chain. The lynx is dependent on the leaves, even though it does not eat leaves itself.

▼ A hungry lynx chases a hare.

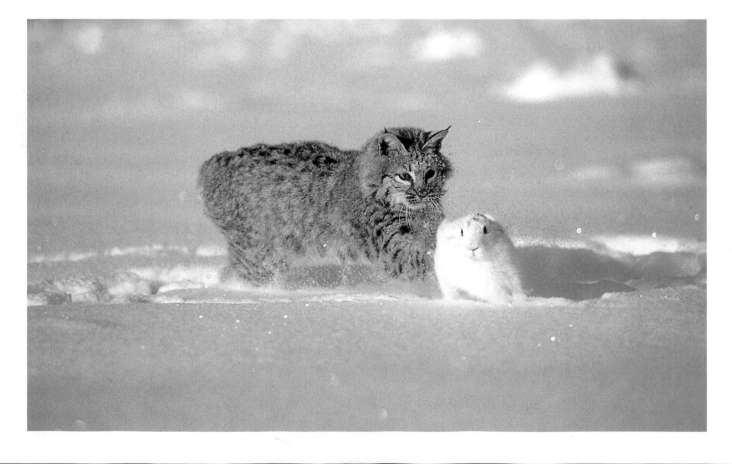

How many habitats?

Within a large habitat, such as a park, there are many smaller habitats, such as leafy bushes or short grass. How many different habitats can you find in your school grounds or nearby park? What kinds of insects or other invertebrates can you find in them? What kinds of birds use them? Do some birds prefer particular trees or bushes?

All living things, including people, need food. They also need oxygen to breathe, as well as water, warmth, and shelter. Each living thing finds these vital resources in its habitat—the environment in which it lives.

A mouse may find food such as ▶ toadstools in its habitat.

Plant food

Plants are green because they contain a green pigment called chlorophyll. This has the ability to trap the energy in sunlight. Plants use this energy in an important set of chemical reactions called photosynthesis. During photosynthesis, carbon dioxide from the air is combined with water inside the leaves to make sugar and oxygen.

Plants use sugar, and the minerals absorbed from the soil through their roots, to make all the other substances they need to grow.

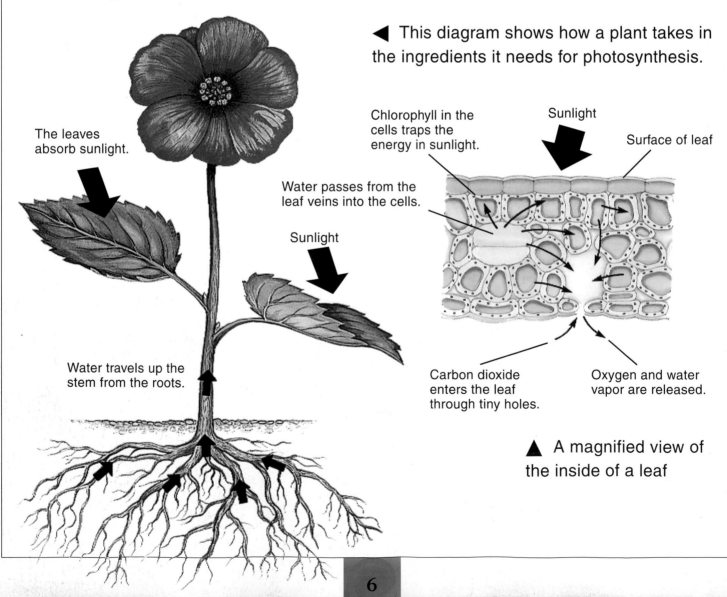

◀ This diagram shows how a plant takes in the ingredients it needs for photosynthesis.

The leaves absorb sunlight.

Chlorophyll in the cells traps the energy in sunlight.

Sunlight

Surface of leaf

Water passes from the leaf veins into the cells.

Sunlight

Water travels up the stem from the roots.

Carbon dioxide enters the leaf through tiny holes.

Oxygen and water vapor are released.

▲ A magnified view of the inside of a leaf

▲ Even water plants, like these water lilies, rely on the sun's energy to make food.

On land, mosses, liverworts, ferns, conifers, and flowering plants grow wherever they can find enough warmth and moisture. Seaweeds and tiny chlorophyll-containing organisms float in the seas. All these photosynthesize, trapping the sun's energy and using it to grow and reproduce. Some energy is stored as starch in swollen roots or stems. Wherever they grow, plants are very important. We could not survive without them.

▼ Vegetables like these provide us with many of the nutrients we need for healthy growth.

Animal food

An animal could sit in the sun all day, but it would not make any food for itself. It cannot turn carbon dioxide and water into sugar—only plants can do this. So all animals depend on plants.

Some animals eat only plants. They are called herbivores. Others eat only other animals. These are carnivores. Some, such as humans, eat both plants and animals and are called omnivores. Scavengers eat the rotting remains of animals left behind by other carnivores, or the bodies of animals that have died naturally.

▼ Scavenging birds called vultures feed on a dead zebra in Tanzania.

Very few animals eat just one kind of plant or prey. If the supply of that one kind of food ran out, they would starve. To ensure that they always have enough to eat, carnivores eat many kinds of other animals, and herbivores munch a wide variety of leaves, fruits, or seeds.

▲ A trout makes a good meal, but otters also eat crabs, frogs, eels, and even young birds.

Herbivore, carnivore, or omnivore?

Make a list of all the pets that you and your classmates own and decide whether they are herbivores, carnivores, or omnivores. Keep a diary of the birds that you see on your school grounds. What do they like to eat? Try putting out separate piles of food such as crumbs, birdseed, grated cheese, or peanuts. Which food does each type of bird prefer?

Energy and Food Chains

Life on earth depends on energy from the sun. Because plants trap and store the sun's energy, they are vital to all other forms of life. Plants are called producers. Animals that get their energy from eating plants are called primary consumers. Animals that get their energy from plant-eating animals are secondary consumers—they receive secondhand the energy trapped by plants.

Producers and ▶ consumers can be arranged in a pyramid diagram like this one.

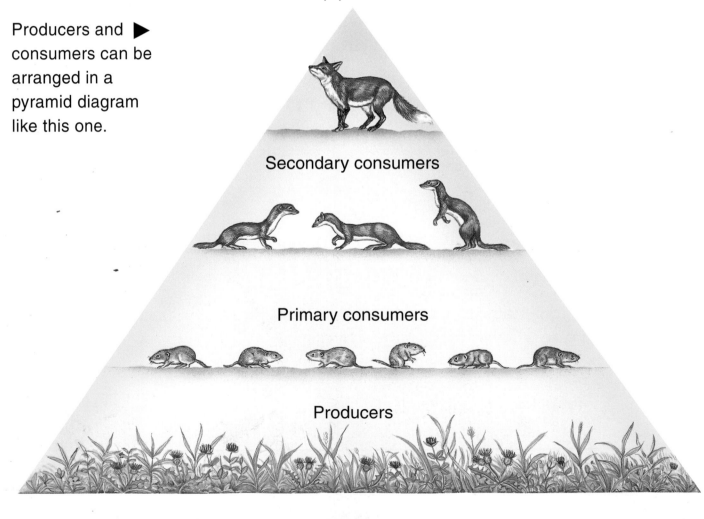

Top predator

Secondary consumers

Primary consumers

Producers

Top predators

At the top of the pyramid of producers and consumers are large carnivores called top predators. Lions, tigers, crocodiles, eagles, and large sharks are all top predators. There are always fewer top predators than the animals on which they prey. If the numbers of prey fall, top predators find it hard to get enough to eat, and so their numbers fall as well.

▼ Lionesses with their prey, a zebra, in Kenya

Food chains

When food is eaten, some of the energy it contains is transferred to the animal that eats it. The rest is lost as heat that escapes into the environment. The transfer of energy from plants to the animals that eat them, and then to their predators, is called a food chain. A food chain is a simple way of showing how food energy goes from plant to predator.

Eaten by a fox (secondary consumer)

↑

Eaten by a rabbit (primary consumer)

↑

Grass (producer)

◄ This diagram shows a short food chain. There are three stages in this chain.

▼ This young fox has caught a ground squirrel.

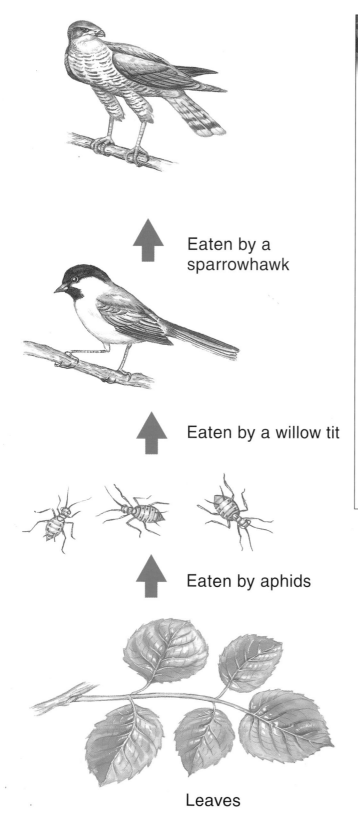

Eaten by a
sparrowhawk

Eaten by a willow tit

Eaten by aphids

Leaves

How many food chains?

What types of plants and animals live in your yard or school grounds? Keep an eye out for insects feeding on leaves or flowers or caught in a spider's web. There might be birds that eat seeds and fruits as well as birds that eat insects. You might spot frogs, mice, and perhaps even chipmunks. Cats, dogs, and foxes might be occasional visitors.

Can you see how all these plants and animals fit into food chains? How many food chains can you think of? Which is the longest chain?

◄ This diagram shows a longer food chain: it has four stages. Sparrowhawks never eat leaves, but they are just as dependent on them as the aphids are.

Food webs

Usually many kinds of plants and animals live close together, and most animals eat all sorts of different things. This is why many food chains are connected to one another. This far more complex network is called a food web. Understanding a food web helps us see how the different kinds of animals depend upon one another and on the plants that grow around them.

▼ This food web shows how plants and animals in an African grassland depend on one another. The arrows point from the food to the animal that eats it.

Lion

Baboon

Zebra

Aardvark

Making a food web

Stand in a circle with all your classmates. One of you names a plant and takes a ball of string and holds the end of it. Then someone names an insect that might eat the plant. He or she takes the ball of string and holds on. Someone else names an animal that might eat the insect, and so on. When someone names a top predator, cut the string and start with a plant again. Can you name more than one creature that might feed on it? How complicated a food web can you make?

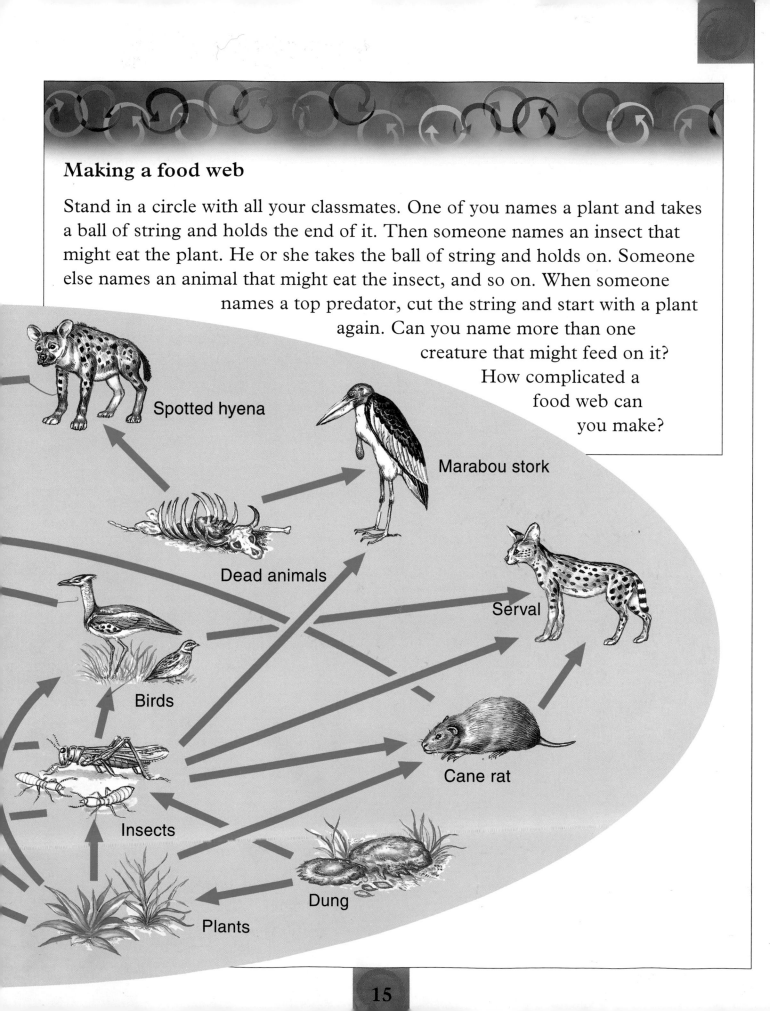

Spotted hyena

Marabou stork

Dead animals

Serval

Birds

Cane rat

Insects

Dung

Plants

Human food webs

For hundreds of years, people ate what they could grow or catch in the area where they lived. For example, the traditional food chain of the Inuit would have looked like the one on the right.

As traders started to carry spices, sugar, and other products to distant lands, diets began to change. Today fast air transportation brings people foods from many different parts of the world, at any time of year, as long as they have the money to buy them.

Polar bear (possibly)

↑

Inuit

↑

Seal

↑

Fish

↑

Photosynthesizing plankton

▼ An Inuit hunter skins a seal. Seals are the Inuit's traditional food.

But there are still differences in the diets of different peoples. The food webs of people living in rural parts of developing countries are mainly made up of foods they can catch or grow themselves. The people of wealthier countries eat much more processed food. Some food and drink seem to be part of the food web almost everywhere. In cities all around the world, you can buy soft drinks, coffee, pizzas, and hamburgers.

Are you a top predator?

Try to make your own food web. Write down everything that you eat in one day, and all the other animals that might also eat these foods if they had the chance. Figure out your place in the food web. Do you think you are a top predator?

▼ This girl's food web includes foods prepared in processing plants and bought from a supermarket.

Productive Habitats

Because plants are the producers, places where lots of plants grow produce the most food and store the most energy. They are known as productive habitats. They can support large numbers of consumers.

Tropical rain forests grow near the equator. Here it is warm all year round, and rain falls almost every day. These forests are always green and leafy. The tall trees are festooned with climbers, and other smaller plants perch upon their branches. The treetops teem with life—insects, birds, snakes, and monkeys. The food web here is very complicated.

▼ Brightly colored macaws (top) and snakes (bottom) make their homes in the rain forests of South America.

In the shallow, sunlit waters off many tropical shores, coral animals grow into large reefs. Tiny green algae photosynthesize and share their sugars with the coral. The reefs are home to brightly colored tropical fish, shellfish, sea urchins, crabs, and octopuses. Reef sharks cruise over them, hunting for prey. Eels lurk in caves in the coral.

There are millions of different kinds of plants, animals, and other living organisms. This huge variety is called biodiversity. The greater the number of different living organisms found in an area, the greater its biodiversity.

▲ Schools of fish dart through the coral reefs off Indonesia.

Less Productive Habitats

In parts of the world where it is very cold or very dry, plants may be few and far between. Little of the sun's energy is trapped and stored. Areas such as deserts, high mountain slopes, and Arctic tundra are some of the least productive habitats in the world. They support far fewer consumers than forests, grasslands, and wetlands.

▼ Mountain goats have to scramble over rocks to find enough plants to feed on.

▲ Caribou living in the tundra move south during the winter to find plants to eat.

Inside the Arctic Circle, summers are short and winters are very long, cold, and dark. Even in summer only the top inch or two of the soil thaws. The rest stays frozen solid. This cold land is called tundra. The plants here grow slowly, making a mat over the ground. It is too cold for trees to grow. Arctic animals have to be tough to survive.

How many plants?

Compare the numbers of plants growing in different places in a yard or a park. Are there as many growing on a weedy path as there are in a lawn or a vegetable patch? Can you think of a way to measure how productive each of these different mini-environments is?

Recyclers

When a plant or an animal dies, the energy it has stored is not lost. Whole armies of scavengers, fungi, and bacteria feed on dead organisms and break down their bodies. These recyclers are called decomposers.

▼ This diagram shows how dead organisms are broken down and recycled.

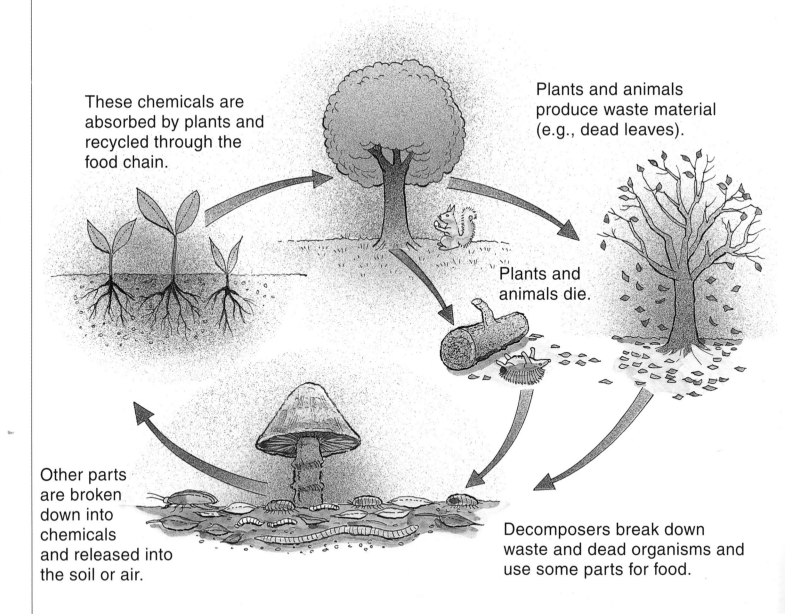

These chemicals are absorbed by plants and recycled through the food chain.

Plants and animals produce waste material (e.g., dead leaves).

Plants and animals die.

Other parts are broken down into chemicals and released into the soil or air.

Decomposers break down waste and dead organisms and use some parts for food.

Slugs, millipedes, and woodlice all chew up bits of leaves, stems, and rotting wood. They produce large, soft droppings that other decomposers feed on. Burying beetles tunnel beneath small dead animals until the bodies disappear below the surface of the soil. Then they lay eggs in the bodies. Earthworms feed on dead leaves, and also on tiny fungi, bacteria, and other remains that are in the soil that they swallow.

▲ A burying beetle gets to work on a dead mouse.

Decomposers at work

You can watch decomposers at work by putting a piece of bread, an apple quarter, half a lemon, and a carrot into separate glass jars. Screw the lids on. Keep a diary to record what you see. As fluffy molds and patches of bacteria grow, the shape of the food changes. Why do you think this is?

◀ Fluffy mold growing on a tomato

Human Interference

Today there are about six billion people in the world. People who study human populations predict that by the year 2050 there may be nearly double this number. We will have to grow more and more food, so there will be even less space for wildlife. Food chains and food webs are already being changed and broken.

◄ Tigers are an endangered species, but they are still killed in places where people think they may be a danger to farm animals.

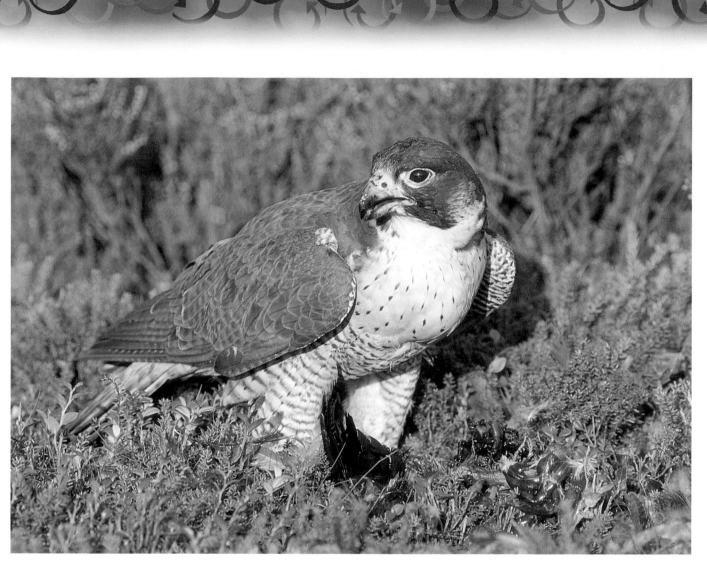

Pesticides

To protect crops from insect pests and diseases, farmers spray them with chemicals called pesticides. Contaminated insects and plants are then eaten by birds and other small animals. The amount of chemicals taken in by these small creatures may not be enough to kill them. But a larger predator may eat several of the small creatures and receive a much bigger total dose. This may be enough to kill it, or prevent it from breeding successfully.

▲ Peregrine falcons were almost wiped out as a result of pesticide poisoning.

Raising animals

In the wild, chickens, pigs, and cattle find their own food. But farmers need their livestock to grow faster and bigger than wild animals, so they give them extra food. Many people are concerned that some food supplements can damage animal and human health. In the 1980s and 1990s, some British cattle developed a disease, called bovine spongiform encephalopathy (BSE), after being given food made from dead sheep, which cattle would not normally eat. Some scientists think people may have developed a similar disease after eating infected beef.

▼ These cattle are eating silage, a food supplement made from grass. Cattle need extra food like this in the winter.

Organic farming

Organic farmers do not use chemicals to kill insect pests and weeds, and they do not give their animals chemicals to make them grow faster. Organic farming methods can help restore food chains by encouraging predators that control pests naturally. Many people think organic foods are healthier, too, because there is no risk that chemicals will be passed from the foods to us.

Natural predators such as ▶ ladybugs feed on aphids that attack crops.

Food from the wild

Although much of our food comes from farms, some still comes from the wild. Sometimes we take too much of this "wild" food and do not leave enough for the natural predators. People have taken so many cod and haddock from the sea that stocks are getting low. Top predators such as seals are often blamed for reducing the fishermen's catches, but really it is people who have ransacked the sea.

Breaking the chains

Play the food web game described on page 15 again. When you have built up a web, imagine that the plant or animal one of you has named has become extinct. That person lets go of the web. What happens to the strength of the web when one organism is lost? What happens to the organisms farther along the chain?

Introducing new species

As people traveled around the world, many plants and animals traveled with them. Some, such as pigs, were transported deliberately, to provide food for people in the lands to which they were traveling. Plant seeds often became caught in cargoes. Animals such as rats were unwelcome companions that infested ships.

These "foreign" species were often released into the wild, either accidentally or on purpose. Often they had no natural predators in their new land. Many of them have multiplied out of control and caused serious problems.

▼ Pigs damage the plant life of some Pacific islands as they search for food.

On remote islands, such as the Kermadec Islands and the Chatham Islands in the Pacific Ocean, the natural food chains were broken when people arrived. They brought with them pigs, sheep, rabbits, and goats that ate too many island plants. Predators such as rats and stoats have disrupted food chains in New Zealand by taking the eggs and young of flightless birds such as the takahe. This bird is now almost extinct.

▲ Rhododendrons brought to Great Britain have spread through woodlands and prevented other plants from growing.

Glossary

algae Organisms containing chlorophyll that live in damp or wet places.

bacteria Organisms made up of a single cell, which can be seen only through a microscope.

chemical reactions Changes that take place when chemicals are mixed together or broken down.

consumers Animals that feed on plants or on other animals.

equator An imaginary line around the middle of the earth.

habitat The natural home of any plant or animal.

invertebrates Animals, such as worms, that do not have backbones.

liverworts A group of green flowerless plants that live in wet places.

native Belonging to a particular place, instead of being transported there.

organism Any living thing.

oxygen A gas that all animals and plants need to survive.

pigment A colored substance.

predator An animal that eats other animals.

prey An animal that is eaten by another animal.

producers Plants that store the sun's energy.

rain forests Thick forests that grow where the weather is wet all year round.

reproduce To produce new plants or animals.

scavengers Animals that feed on dead organisms.

silage An animal food produced from crops such as grass, cut while it is still green and juicy.

Further Information

BOOKS

Ganeri, Anita. *The Hunt for Food.* Brookfield, CT: Millbrook Press, 1997.

Parker, Steve. *Food & Digestion.* (Human Body.) Danbury, CT: Franklin Watts, 1990.

Pirotta, Saviour. *Predators in the Rain Forest. Austin, TX: Raintree Steck-Vaughn*, 1998.

Riley, Peter. *Straightforward Science: Food Chains.* Danbury, CT: Franklin Watts, 1998.

WEB SITES

http://www.greenpeace.org
Contains information about the environmental organization Greenpeace and its activities.

http://www.wwf.org
The World Wildlife Fund's home page, with information about the work of WWF and wild places all over the world.

http://www.envirolink.org/
The Envirolink page, with information on a wide range of environmental issues and links to other useful sources of information.

http://www.gn.apc.org
The home page of GreenNet, a computer network specializing in environmental issues.

Index

Page numbers printed in **bold** mean that there is information about this topic in a photograph, diagram or caption.

© Copyright 2000 Hodder Wayland